Introduction to Hadoop for Bioinformatics

Martin Gollery

Tahoe Informatics

ISBN:1508464464
ISBN-13:9781508464464

DEDICATION

This book is dedicated to those who wish to make the discoveries that will make a difference in the World. This book is designed to provide a brief introduction to Hadoop, which may be seen as the picks and shovels for the gold mining of the 21st century- the mining of biological information.

"Hiding within those mounds of data is knowledge that could change the life of a patient or change the world" -Atul Butte

CONTENTS

ACKNOWLEDGMENTS

I would like to acknowledge my daughter Sharon Gollery for her help with this document, my wife Suzanne for her support and guidance over the last 25 years, and most importantly, my students who inspired me to compile these notes into this short book!

INTRODUCTION

Hadoop

Why should people use Hadoop? Haven't people been using clusters for decades?

Well, yes they have, but the Grid Engine technology has grown old. The Grid technology was a great leap forward over the earlier mainframe technologies that preceded it, but it has grown tired. Storage clusters have the burden of lugging data back and forth across the network, and require higher and higher bandwidth and controller technologies. Still, those who boast of thousands of cores have often found that those cores were poorly utilized. Increasing the number of processing nodes may move the bottleneck, but it does not eliminate it. Scaling the system frequently involves reindexing data, rebuilding system configurations, and unpleasant downtime.

Hadoop breaks a problem into pieces and spreads them across the cluster. Storage is done locally, with no expensive RAID systems and the rebuilding processes that seem to take forever. Replication is done by the system, and scaling that system is vastly easier than with older technology.

Hadoop for Bioinformatics

Hadoop is now practically synonymous with big data. The amount of data that is now being generated by Next-Generation Sequencers is considered to have the Velocity, Volume and Variability to qualify as 'Big Data'. Therefore, it behooves us to move towards tools that were designed to handle this type of work.

After all, you might be able to cut down a tree with a crosscut saw, but if you want to cut down a thousand trees to stop an oncoming forest fire, you need a lot of people with a lot of chainsaws- big ones!

Hadoop in Bioinformatics is still at the early stages. The purpose of this book is to provide a brief introduction, in simple terms, to the Hadoop ecosystem, what all those strange names mean, and how it all fits together.

But what about Accelerators?

It is true that I have worked with accelerated Biocomputing for many years- FPGA's, GPU's, ASIC's, Phi… so why would I switch to Hadoop?

The answer is… I didn't. Hadoop spreads the problem across some number of nodes. Accelerators work better with Hadoop, in fact! This is because the accelerators are frequently starved in most systems- the system can't feed them data fast enough!

With Hadoop, the files are broken up and distributed. The storage is local, and the pieces are smaller, therefore, the I/O bound processes are no longer so much of a restriction, and the accelerators are free to run.

Tahoe Informatics

I am the owner of Tahoe Informatics, a Life Science consulting firm based in Lake Tahoe, Nevada. After working in the Oil and Aerospace industries, and reaching at the High School and college levels, I have been in Bioinformatics since 1998. Specializing in Acceleration Technologies, I have published in a number of journals and presented at conferences worldwide. I consult with companies and academic institutions either remotely or in person to solve problems with data analysis, training, fund raising (through grants and venture capital), and marketing.

HADOOP HISTORY

Google published papers on the Google File System and the Map/reduce system in 2003/2004. Hadoop was created using concepts outlined in these papers by Doug Cutting and Mike Cafarella in 2005. Cutting, who was working at Yahoo! at the time, named it after his son's toy elephant. It was originally developed to support distribution for the Nutch search engine project.

Hadoop was later spun out of Yahoo! and is now a project of the Apache Software Foundation.

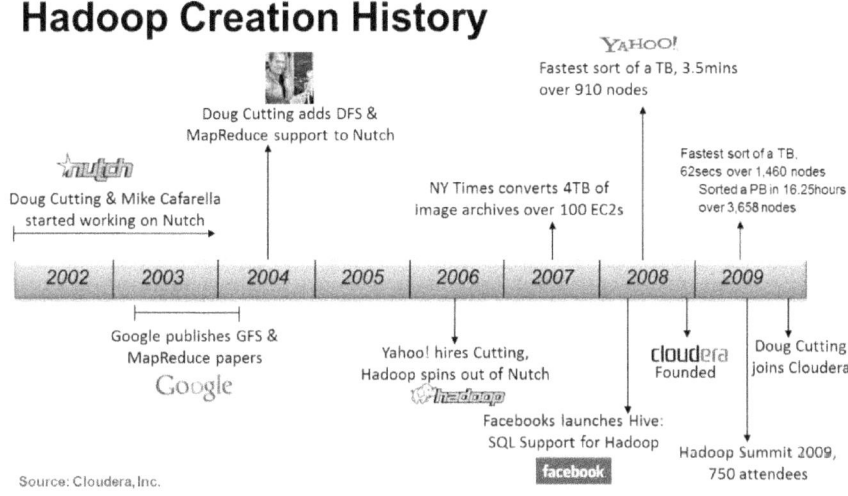

WHAT IS HADOOP?

Hadoop is a framework of tools for running applications on Big Data. It is an open-source Apache project, which means that you can get it for free. You can also buy it from several companies like Cloudera, Hortonworks, MapR or others.

Wait, why would you buy software that you can get for free?

Well, the corporate software has additional features- management, installation, training, extensions and most of all… support! So, people will spend money on Hadoop. Whether it is worth it to you depends on your own situation. Can you support it yourself with your own IT staff? Do you need the additional features? These are the kinds of questions you need to ask before making the decision.

Hadoop is available:
-On Linux
-On Windows
-On cloud

You can also use a 'Sandbox' version of Hadoop so that you can play around with it a bit before you commit to buying a big cluster and spending the next 6 months setting it up!

For a small test cluster, some of the micro-instances on the Amazon Cloud are a good deal. A 10-node cluster of these systems comes to less than 15 cents /hour, enabling you to really get a feel for how this works without a big outlay of cash.

WHO USES HADOOP?

Yahoo
FB
Ebay
Twitter
LinkedIn
About half the big companies in the world.

A few years ago, Yahoo said it had a cluster of 42,000 Hadoop nodes. Others have systems with tens of thousands of nodes. Facebook grows by over half a Petabyte per day! Moreover, FaceBook was able to move their entire operation from one location to another, with no downtime!

Why do we care what these big companies do? Well, they are ahead of life sciences in the processing of Big Data. It behooves us to learn from what they do, and they are creating a roadmap that we can follow.

The best thing about open source software is that these companies have thousands of developers that improve, extend, simplify and document all of the tools in the Hadoop ecosystem. The pace of progress is unbelievable, and we users are the beneficiaries!

WHY USE HADOOP?

Hadoop is scalable, as I showed before. Expanding a Hadoop cluster is much easier than legacy systems. Since the storage and CPU are in the same box, adding a new rack of nodes to the system increases the speed of the entire system, rather than simply moving the bottleneck around.

Hadoop is cost-effective because it uses commodity hardware. You don't need expensive RAID controllers because Hadoop has the redundancy built in for the entire system, not just the drives. You don't need supercomputer-class network connections because less data is being transferred across the network. You don't need the highest reliability drives and components because a node failure in Hadoop is not a big deal.

Hadoop is flexible because you can reassign and reallocate nodes more easily than in other systems.

Hadoop is fault-tolerant because it is designed to be so from the ground up. The system assumes that node failures are expected and common. Data is triplicated, so in the event of a failure, two more copies are still available.

Hadoop is fast. It has performed a number of benchmarks much more quickly than other systems.

HOW DOES HADOOP SCALE?

While Hadoop clusters have been reportedly built with over 42,000 nodes, it is likely that much bigger ones have been built since then. I have heard rumors that some clusters exceed 100,000 nodes, but have not verified.

Storage systems have gone well over 100 PB.

Some clusters process over 500 TB per day. Even a roomful of the fastest sequencers can't produce this much!

Note that these are old numbers- probably much higher now!

As a result, Hadoop is nearly synonymous with 'Big Data'

"Hiding within those mounds of data is knowledge that could change the life of a patient or change the world" -Atul Butte

THE HADOOP ECOSYSTEM

THE HADOOP DISTRIBUTED FILE SYSTEM

The Hadoop Distributed File System- HDFS. This was an improvement on the original Google File System.
Data is triplicated across HDFS, eliminating need for expensive RAID controllers. Note that some people prefer other file systems for various reasons. HDFS can be used with others- like Lustre from Cray, which sits on top of HDFS.

How does HDFS work?
  -Take a big file
  -Split it into blocks- the size can be configured.
  -Create 3 replicas
  -Distribute those replicas across the cluster
  -If a node goes down, use copies mapped to other nodes
  -Process data on each node

MAP/REDUCE

Should really be called Map/Shuffle/Reduce. This is the basic framework for parallel processing.
Provides scalability & fault tolerance
Keys & values
Extract data you care about- Map to processing nodes. Each worker node applies the "map" function to the local data, and writes the output to a temporary storage. A master node orchestrates that for redundant copies of input data, only one is processed.

Shuffle- Distribute sorted Map output to reducers. Worker nodes redistribute data based on the output keys (produced by the "map()" function), such that all data belonging to one key is located on the same worker node.

Reduce step aggregates, summarizes, outputs results. Worker nodes process each group of output data, per key, in parallel.

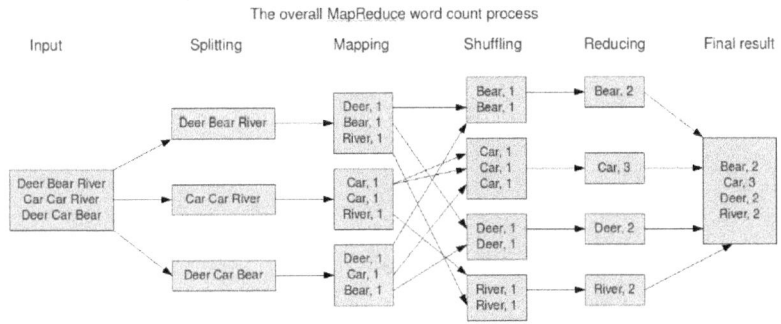

http://www.rabidgremlin.com/data20/

PIG

Works with ETL type tasks- Extract, Transform, Load. These tasks are scripted in a language called "Pig Latin". So, Transformations such as aggregate, Join and Sort are written as a SQL-like script. Pig Latin is different than SQL in that it is a flow language, rather than a declarative language. SQL is best for asking a question of the data, but Pig Latin allows you to specify a data flow that describes how your data will be transformed. Both have their uses in the analysis of Life Science data.

In general, Pig may be viewed as a High Level tool for Map/Reduce. Pig is extendable with other languages such as Python, Java, etc.

Why use Pig?

Java Map/Reduce jobs are too painful to write, and require greater knowledge. You don't need to be a programmer to write Pig scripts, and even if you are, Pig can save you 95% of the time required to do the work. If you are a programmer, you can create your own User Defined Functions (UDFs) which can then be called from Pig.

Pig jobs are Self-Optimizing, so the user can focus more on the task at hand, not on IT work.

Google developed a similar tool called Sawzall.

HIVE

Hive is a data warehouse that can be queried with HQL, similar to SQL. Hive was developed by FaceBook, now used by Netflix, etc. More SQL-like than Pig Latin, so Hive requires some sort of Schema. The query is automatically converted into a Map/Reduce, Tez or Spark job.

Hive supports indexing, compression, User-defined functions.

Hive was originally Batch oriented, and replies could take several minutes to return. Recent initiatives have brought it down into the sub-second range, even with huge datasets. Still, you might want to consider Impala.

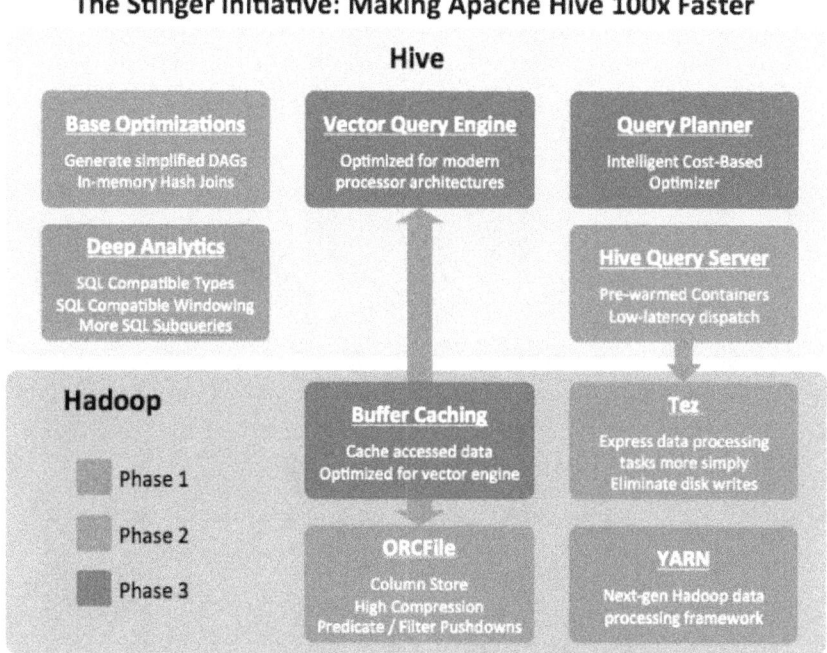

https://gigaom.com/2013/04/30/with-impala-now-ga-clouderas-ceo-sizes-up-the-sql-on-hadoop-market/

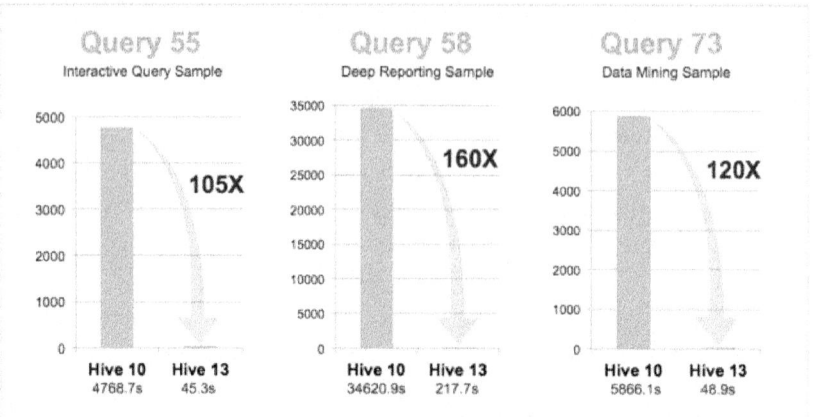

http://blog.newitfarmer.com/category/anls/sql-on-hadoop/hive-bi

YARN

Yet Another Resource Negotiator

YARN is the key difference between Hadoop 1 and Hadoop 2. YARN manages resources, schedules applications, and allows diverse processes. As a result, applications that are not 'Hadoop Optimized' can still be run on a Hadoop cluster.

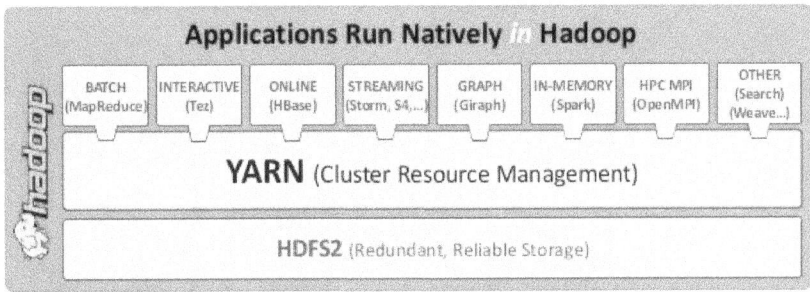

(Hortonworks.com/Hadoop/yarn)

Allows many other apps besides MapReduce.
Improved cluster utilization
Multiple versions of same app allowed- very important for NGS workflows!

YARN is
-Flexible in allowing multiple applications, and multiple versions of the same app.
-Efficient in improving cluster utilization
-Scalable in removing the limits on cluster size
-Share same cluster across applications

YARN Speed Benefits

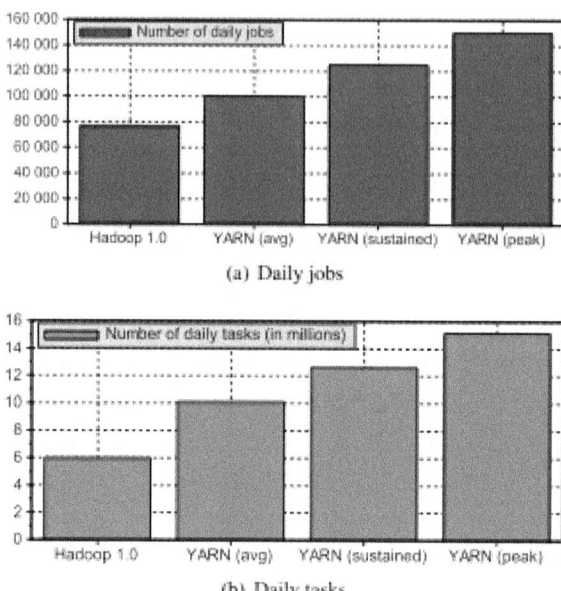

(a) Daily jobs

(b) Daily tasks

Figure 2: YARN vs Hadoop 1.0 running on a 2500 nodes production grid at Yahoo!.

TEZ

Tez runs on top of YARN. A Tez job is expressed as a Directed Acyclic Graph (DAG). Data is processed as a task, which becomes a node in that graph. When a task is complete, the data is picked up by the next node.

Tez is Data type agnostic- as long as the data can be processed by the application that makes up the next node, it can be run under Tez. Hive and Pig queries that would require many M/R jobs can run as a single Tez job.

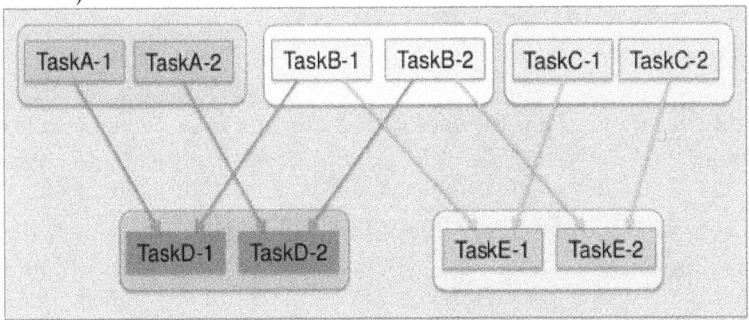

(Murhty/Saha Hadoop Summit)

Tasks in this example might be- 1. Download SRA file 2. Convert to FastQ 3. Deduplicate 4. Trim 5. Map to Genome 6. Convert SAM to BAM

HBASE

HBase is a Column oriented, non-relational database that is distributed on HDFS. It compresses data, and operates in-memory. Speed is much better in later versions than when it first came out, but it still is not a real replacement for a SQL database.

HBase tables can serve as input/output for Map/Reduce jobs. So, if you have data in a table that you want to process in some way-trimming or normalizing, for example- then the table can be easily processed across the cluster.

Good for sparse data, big tables that are not fully populated, such as variants. HBase is very flexible, so you don't need to have structured data.

Performance is quite linear. Doubling the nodes pretty much doubles performance.

HBase is highly scalable- Stumbleupon claims that their system has 9 Billion rows, 1.3 PetaBytes.

How fast is it? You can Mapreduce a 700GB table in 20 minutes. Companies such as Facebook, Twitter, Yahoo, and Adobe use HBase internally. The facebook messaging platform, for example, uses HBase.

HBase Architecture

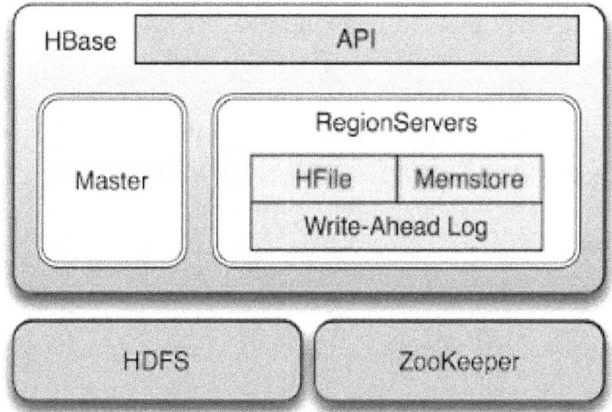

http://bigdatariding.blogspot.com/2013/12/hbase-architecture.html

HBase Tables:

Tables are sorted by Row in lexicographical order
Table schema only defines its column families
Each family consists of any number of columns
Each column consists of any number of versions
Columns only exist when inserted, NULLs are free
Columns within a family are sorted and stored together
Everything except table names are byte[]
Hbase Table format (Row, Family:Column, Timestamp) -> Value

Hbase consists of:

Java API, Gateway for REST, Thrift, Avro
Master manages cluster
RegionServer manage data
ZooKeeper is used the "neural network" and coordinates cluster
Data is stored in memory and flushed to disk on regular intervals or based on size
Small flushes are merged in the background to keep number of files small
Reads read memory stores first and then disk based files second
Deletes are handled with "tombstone" markers

DRILL

Drill provides real-time interactive analysis of Big Datasets. Drill is the open-source version of Google's 'Dremel' tool.

Drill uses schema-free queries- think exploratory searches rather that 'What is coverage at position 23,391 on chromosome 11?'

Blazing Fast!

Drill is able to scale to 10,000 servers or more and to be able to process petabytes of data and trillions of records in seconds.

GIRAPH

Graph database for Hadoop. Utilizes Map/Reduce to process the Graphs.

Giraph is built for high scalability – Giraph was used by FaceBook to analyze one trillion edges in 4 minutes, using only 200 machines! They showed that the performance scaled almost linearly when more workers were added.

Network Analysis
Clinical Trial data
K-means Clustering

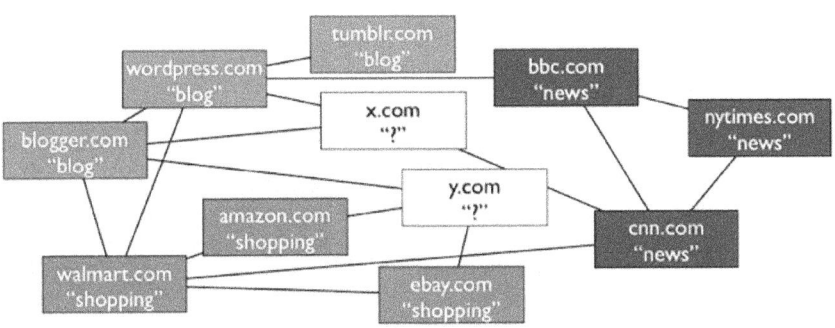

"An example of label propagation: Inferring unknown website classifications from known website classifications in a graph where links are generated from overlapping website keywords."
-FaceBook Engineering report

So, FaceBook showed website classification inferences. In Life Sciences, we can use this same technique to infer classifications that matter to us- Genes, Interactions, Promoters, or Drug Targets.

IMPALA

SQL Query Engine for Hadoop- From Cloudera, but open source. Similar to Drill and Dremel, in that it is very low latency.

Impala is massively parallel- as a result, it is up to 100x faster than Hive. Data processing and interactive queries can be done using the same data & metadata without having to move it to a special system and special formats just to perform analysis.

Great for Exploratory Analytics with Previously Unknown queries
Queries on big & growing datasets
EDW/RDBMS can't:
> Dump in data, then later define schema/query
> Evolve schemas easily
> Scale by just adding nodes
> Store cheaply

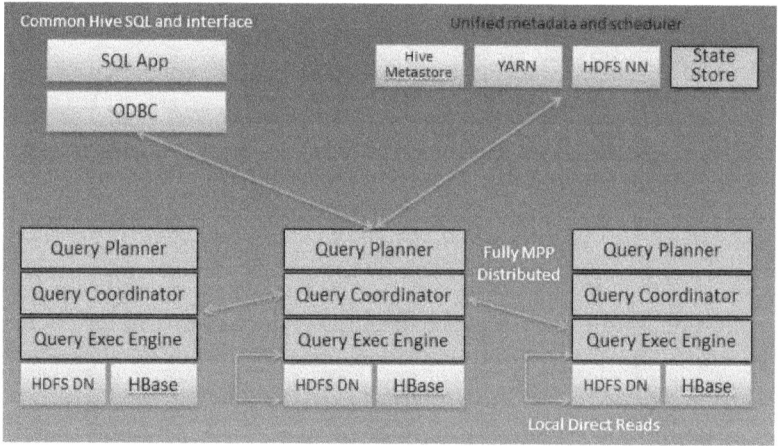

(http://www.theregister.co.uk/2012/10/24/cloudera_hadoop_impal a_real_time_query/)

SOLR

SOLR is an enterprise search engine for all the papers, whitepapers, online textbooks, lab reports, emails, etcetera, that your organization has in house. SOLR is for document processing- not DNA analysis.

Documents are fed in over HTTP and indexed by SOLR.
Query SOLR with HTTP GET command. You can do full text searching with phrases, wildcards, joins, groupings.

SOLR can use Schema or be Schemaless
Geospatial search- imagine searching data only from SE Asia, for example
Faceting features can group and organize data in a multitude of ways
Data types can be in JSON, XML, Word, PDF, excel, etc.
"Shard splitting" means that it is easy to add capacity without re-indexing everything. Solr scales to hundreds of thousands of nodes, billions of docs

Lucene/Solr Architecture

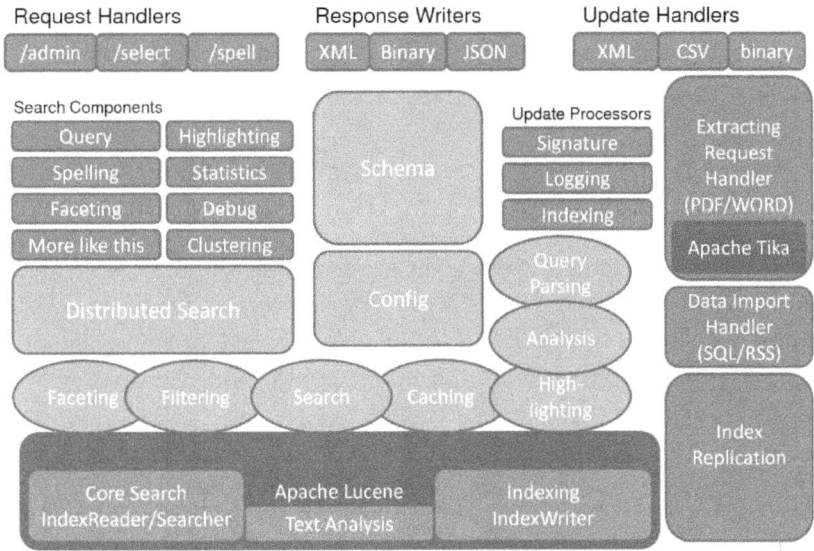

http://www.sunilgulabani.com/2013/01/apache-solr-introduction.html

MAHOUT

Mahout is a scalable Machine Learning framework (ML)

What can we use Mahout or other ML algorithms for in the life sciences?

-Filtering
-Clustering
-Classification

Mahout is unusual compared to most of the other frameworks that we have been discussing, in that it can also be used on a single node, without Hadoop. So, if this is the only tool you need for a project, you do not need to build a whole Hadoop cluster to do it.

You probably already use Mahout.

The recommendation engines for Netflix, Pandora, etc. are based on Mahout. So, when you see something that says you might also like a movie, it is because it is using the data that it already knows about you- people that like Star Wars, 2001, and Doctor Who have a high probability of also liking Star Trek.

Recommendations are based on classification. Classification uses known data to classify unknown data.

Clustering uses characteristics that are found in the data, not known groups. For example, genes that are expressed more in certain situations, such as drought stress in plants. Clustering may be done using expression levels in these stress conditions without knowing that they are associated.

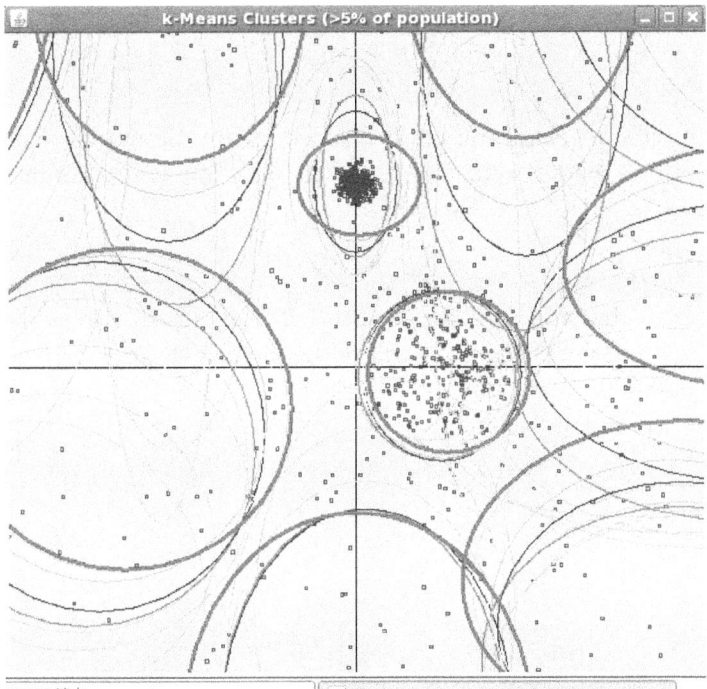

http://cdn-ak.f.st-
hatena.com/images/fotolife/y/yutakikuchi/20120503/20120503125606
.png

STORM

Storm is a distributed real-time computation system. Streaming data processing, rather than Batch. Can be used with any programming language.

Very fast- 1,000,000 tuples/Second/node
Mining Tweets, other social media for information. Storm is reliable, and guarantees that every message will be handled. Twitter uses it to process 400,000,000 tweets/day.
Used by WebMD medpulse

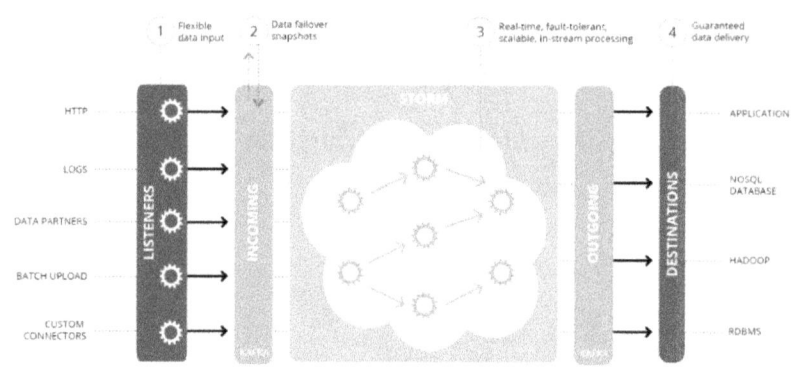

Infochimps.com

SPARK

In-memory
2-10x faster even for on-disk tasks
Up to 100x faster than Hadoop MR
Works with Hadoop's storage API's
HDFS, Hbase, S3, etc
Many operators are available right out of the box, so development is easier.

Spark Speed

Spark

What kind of iterative questions?
How about loading up 1000 BAM files, then querying coverage at various points/conditions?

SparkR

R jobs at scale
From UC Berkeley- started as a class project!

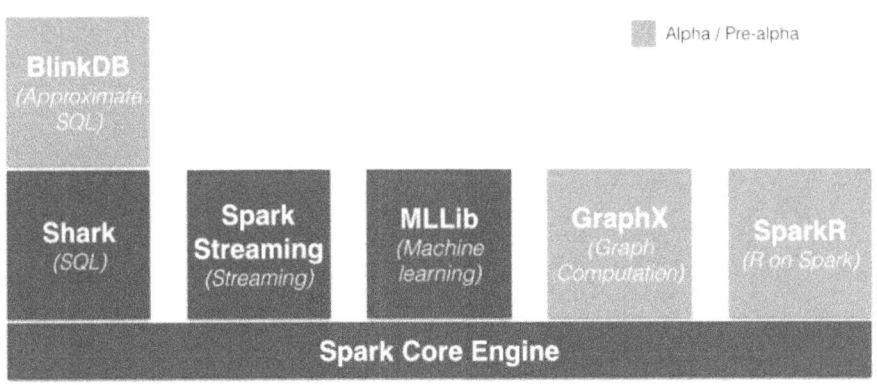

The Ecosystem of Spark Projects (DataBricks)

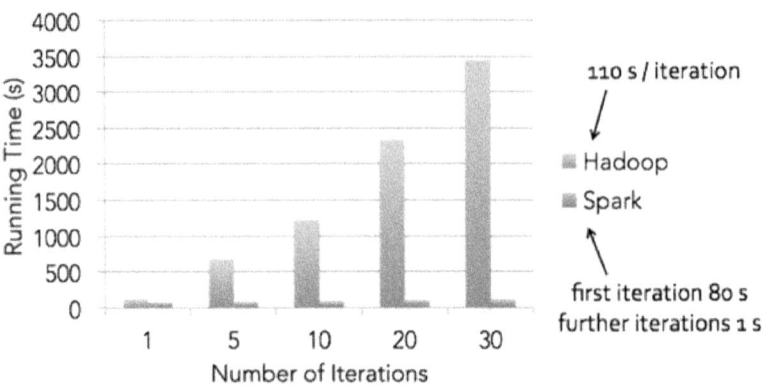

(Cloudera.com)

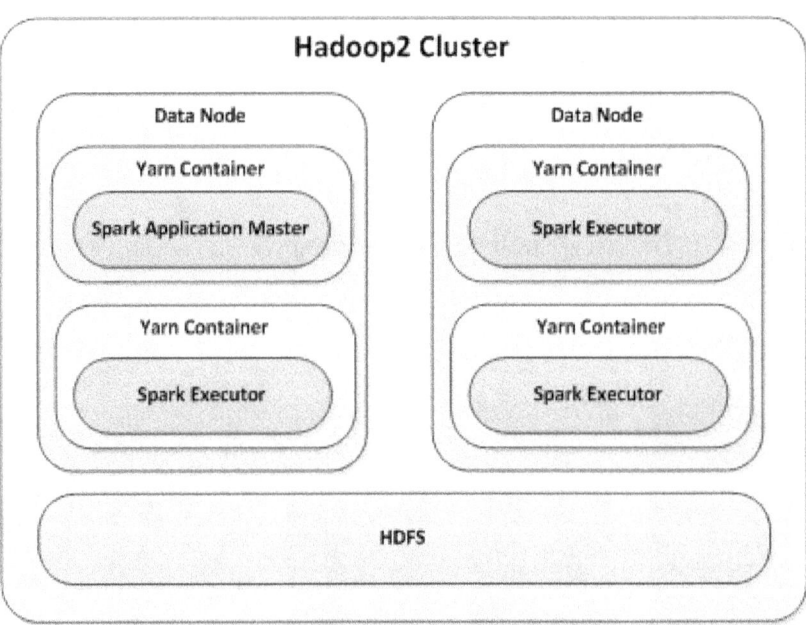

Spark masters and executors sit inside YARN containers on the data nodes. (ebaytechblog.com)

SQOOP

Pronounced 'Scoop', Sqoop sits between relational databases and Hadoop.

Parallelizes data transfer- you Sqoop the data from Oracle to a Hadoop resource, for example. Sqoop copies data quickly into HDFS, Hive, Avro or Hbase

Sqoop is flexible!

Sqoop works with relational databases such as: Teradata, Netezza, Oracle, MySQL, Postgres, and HSQLDB.

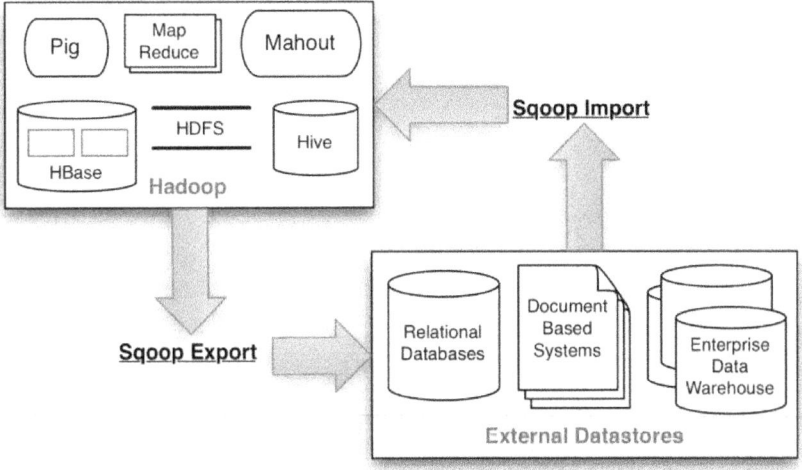

The Sqoop import/Export process (http://divakarbigdata.blogspot.com/)

OOZIE

Workflow scheduler system to manage Apache Hadoop jobs.

Oozie <u>Workflow</u> jobs are Directed Acyclical Graphs (DAGs) of actions.

Oozie <u>Coordinator</u> jobs are recurrent Oozie Workflow jobs triggered by time (frequency) and data availabilty.

Oozie is integrated with the rest of Hadoop supporting several types of Hadoop jobs out of the box (such as Java map-reduce, Streaming map-reduce, Pig, Hive, Sqoop and Distcp) as well as system specific jobs (such as Java programs and shell scripts).

Oozie is a scalable, reliable and extensible system.

Oozie

Oozie triggers workflow actions, but Hadoop MapReduce executes them.

Oozie detects completion of tasks through callback and polling. When Oozie starts a task, it provides a unique callback HTTP URL to the task, thereby notifying that URL when it's complete. If the task fails to invoke the callback URL, Oozie can poll the task for completion.

Oozie Coordinator can also manage multiple workflows called a "data application pipeline".

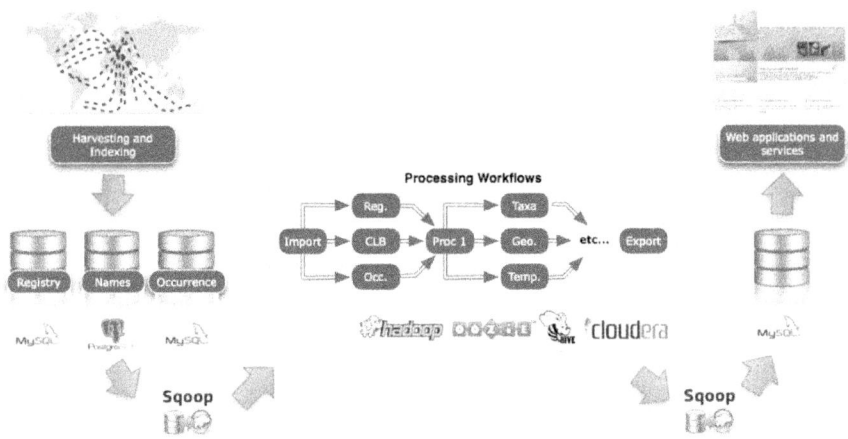

Oozie is the service that manages the workflow for Hadoop (blog.cloudera.com)

FLUME

Designed to flow data from a source into Hadoop. So far, I have only seen it used for web logs, twitter feeds.

Agents are spread through the IT infrastructure – inside web servers, app servers and mobile devices, for example – to collect data and bring it into Hadoop.

Could be used for EHR data- imagine nurses/doctors with tablets. Once the data is incorporated into the system I can be combined with other sources, and mined with Solr.

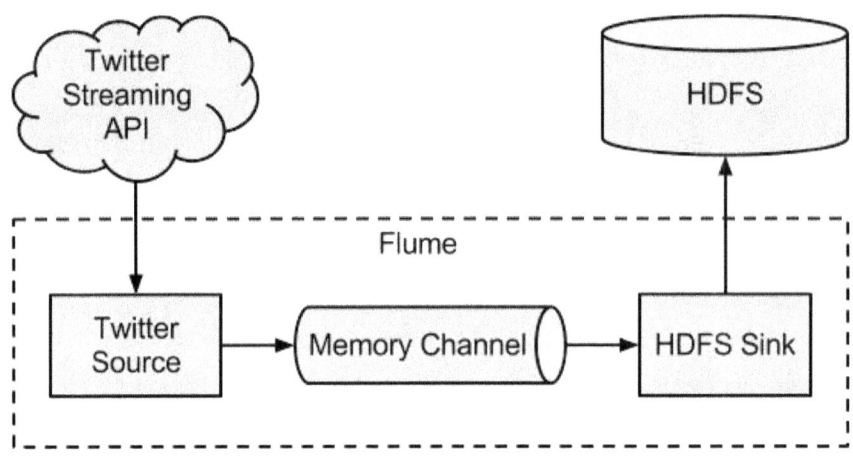

Blog.cloudera.com

ZOOKEEPER

'One Ring to rule them all, and in the darkness, bind them!'

ZooKeeper is a centralized service for maintaining configuration information, naming, providing distributed synchronization, and providing group services. All of these kinds of services are used in some form or another by distributed applications. Each time they are implemented there is a lot of work that goes into fixing the bugs and race conditions that are inevitable. Because of the difficulty of implementing these kinds of services, applications initially usually skimp on them ,which make them brittle in the presence of change and difficult to manage. Even when done correctly, different implementations of these services lead to management complexity when the applications are deployed.

The main differences between ZooKeeper and standard file systems are that every znode can have data associated with it (every file can also be a directory and vice-versa) and znodes are limited to the amount of data that they can have. ZooKeeper was designed to store coordination data: status information, configuration, location information, etc. This kind of meta-information is usually measured in kilobytes, if not bytes. ZooKeeper has a built-in sanity check of 1M, to prevent it from being used as a large data store, but in general it is used to store much smaller pieces of data.

HedWig
>   Publish/subscribe model
>   Large amounts of data across the Internet
>   Guaranteed delivery
>   Metadata on topics, subscribers and hubs is kept in Zookeeper.

Bookeeper- related system, manages log streams from nodes
>   Servers (bookies) send log streams (ledgers) and it is tolerant of failures.
>   Metadata is sent to Zookeeper

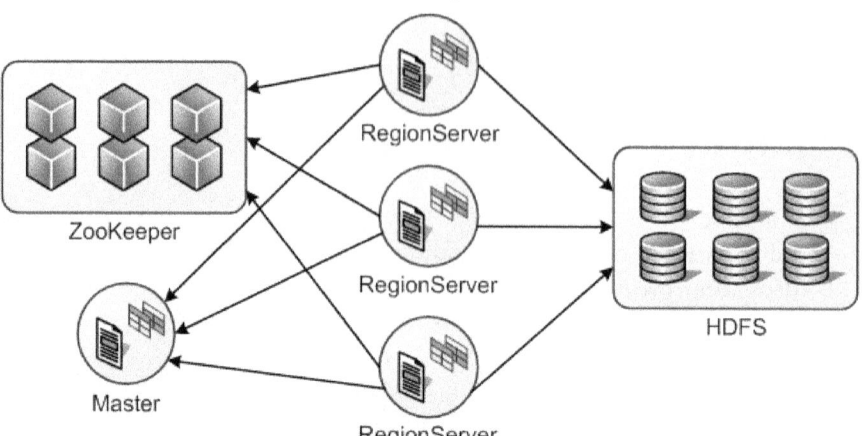

HADOOP IN THE LIFE SCIENCES

Now that we have looked at the components of the Hadoop ecosystem, we are going to take a look at Hadoop implementations.

Commercial
Cloud
Hardware
Academic

Commercial Hadoop

Hadoop is an Open-Source system, managed by the Apache Foundation. So why would you want to buy it? Well, a number of features can be added, such as management tools, and of course, if you buy a commercial version then you can get support.

The concept is similar to Linux. While many people will simply download Linux for free, many will also buy the official RedHat version.

Below are some of the popular Hadoop vendors:

HortonWorks makes the 'HortonWorks Data Platform Powered by Hadoop'. This is 100% open source. HortonWorks will also provide Technical Support, Training and Partner services.

Cloudera is the biggest Hadoop vendor. Intel was working on their own version of Hadoop, but decided instead to join in with Cloudera (along with a very large investment). They provide training, architectural services, and technical support.

Various companies, such as Dell, will get certified to work with Cloudera, which gives one the confidence that a cluster built with those components will work with Cloudera.

IBM has the BigInsights platform, which extends Hadoop with Advanced Analytics, performance optimizations, workload management, Visualization tools and security.

<u>MapR</u> sells a higher performance Map/Reduce system, with better fault tolerance, and read/write access to the file system via NFS.

Pivotal had their own distribution which has now been open sourced. They have a parallel SQL processing engine called HAWQ and a real-time analytics engine called Gemfire XD that runs in memory.

Hadoop Hardware

Oracle BDA (Big Data Appliance) includes Cloudera Hadoop, Impala and the Oracle NoSQL database. This system also includes the Oracle R distribution, which is many times faster than standard R- up to 170 times faster in matrix multiplication tasks!

Cray Urika-XA comes with Cloudera pre-installed. The Cray Lustre file system sits on top of HDFS, and the performance is enhanced by Infiniband connections, SSD storage, and Sonexion systems.

Teradata has partnerships with Cloudera and HortonWorks running on their appliances and with their software.

Dell, HP, SGI, etc. all claim to be compatible with Hadoop, and most probably are. You can easily check to see if they are certified by Cloudera,

Hadoop in the Clouds

The advantage of running Hadoop in a cloud service is that no upfront cost is necessary, and a very large cluster can be provisioned in a very short time. Projects can be prototyped and concepts verified easily, which can save a great deal of time and money.

In production mode, cloud based systems can be very effective when needs fluctuate, as they often do. If needs are constant 24/7, then the economics are usually in favor of a local installation.

AWS
>   EMR- 10 node cluster starts at 0.15/hr
>   1700 genomes from the 1000 genome project-already there!

Google
>   Run Yarn, Spark from the Developers console on Google Compute engine
>   Hortonworks or MapR

MS Azure
>   HDinsight
>   Integrated with Hortonworks for cloudbursting
>   Query on-premises and cloud data at the same time

Life Science Software-Commercial

BioDT (BioDatomics.com) Provides a workflow development tool that includes ~250 bioinformatics tools. Designed to work with Cloudera Hadoop.

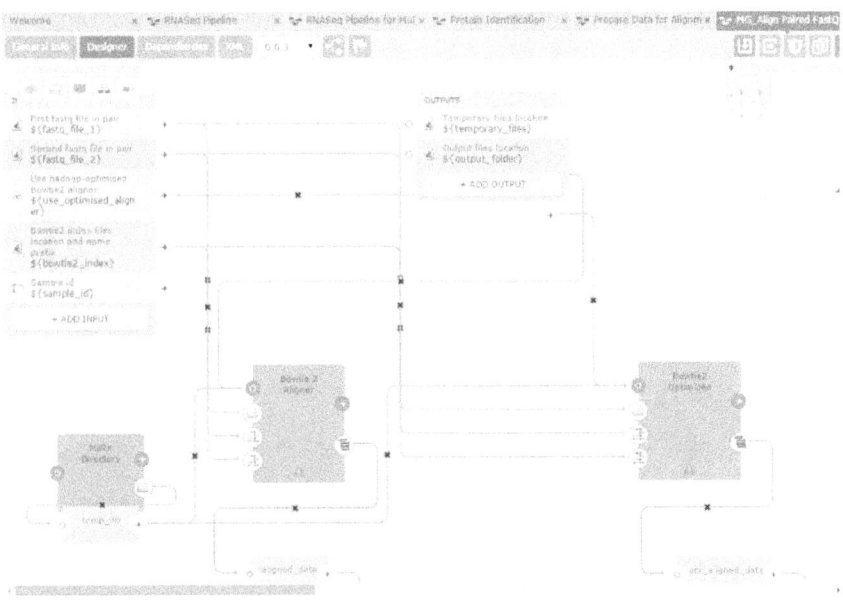

RapidMiner works with Hadoop to form 'Radoop', and provide Advanced Analytics. RapidMiner is the most widely used platform for Advanced Analytics, and has been used in GWAS studies.

Rapidminer partners with both Cloudera and HortonWorks.

SAS is a well-known Statistics and Data Mining company, that collaborates with HortonWorks and Cloudera to add analytics power to Hadoop.

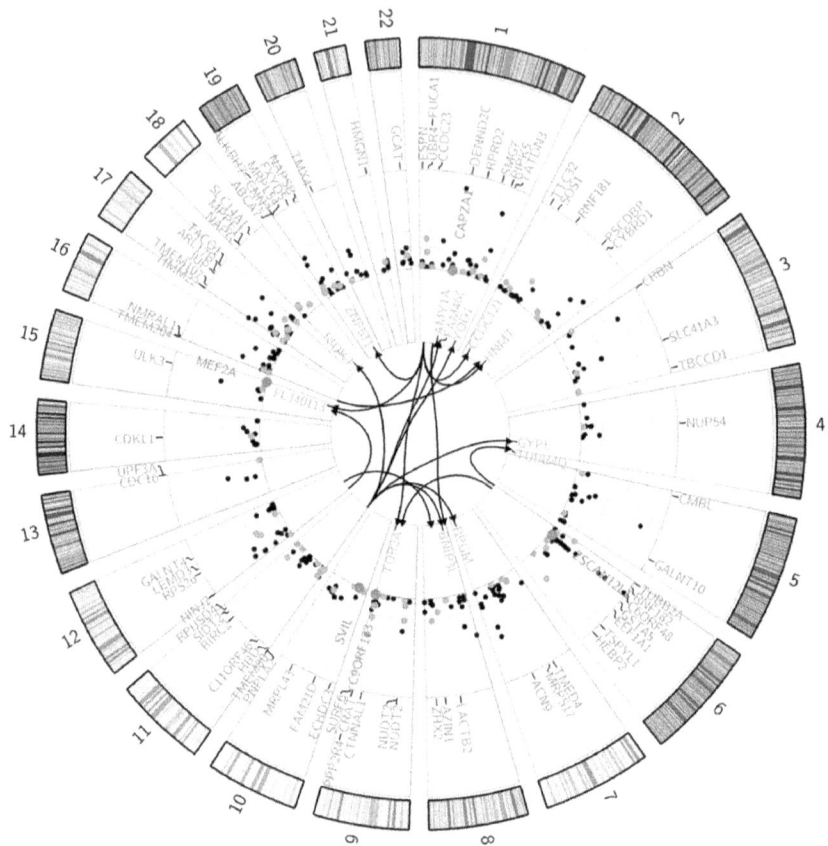

Genetic regulation of gene expression in SCD patients.
(http://journal.frontiersin.org/Journal/10.3389/fgene.2014.00026/full)

Academic Endeavors

Global Alliance for Genomics and Health

242 members- Public & Private

The Genomics API has been modeled with Apache Avro

Will standardize on these new technologies:
"current sequence processing pipelines must be redesigned to leverage existing frameworks, such as MapReduce, Hadoop, and Spark"

ADAM

ADAM is a good example of a new Hadoop-based standard that is associated with the Global Alliance. ADAM is seen as the successor to SAM and BAM. It builds on Three main features:

Spark- Which we talked about earlier.

Avro- a data serialization system

Parquet- a very efficient columnar data storage system. Parquet supports very efficient compression and storage techniques. It works with all the other Hadoop tools, so data stored in ADAM is easy to get at.

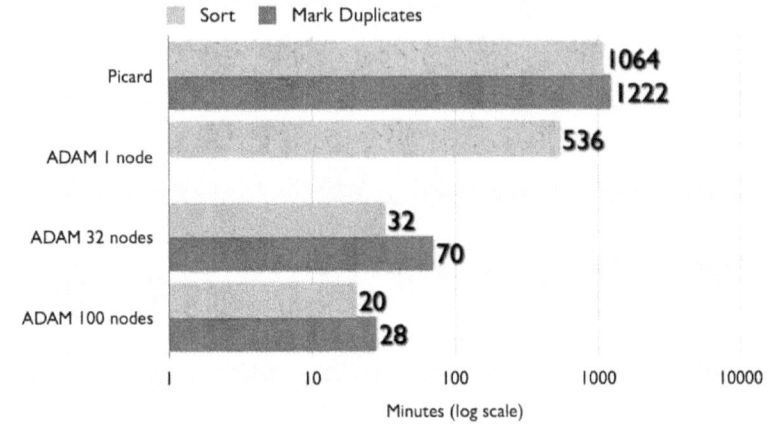

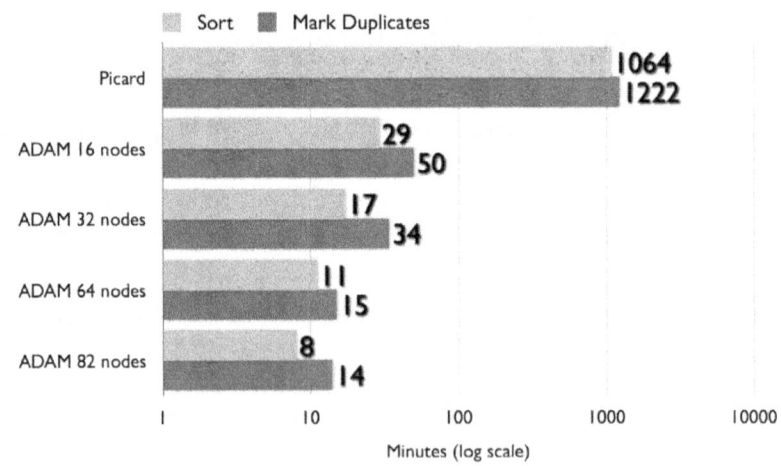

http://www.slideshare.net/mattmassie/strata-big-data-science-talk-on-snap-adam-and-avocado

CROSSBOW

Map: use Bowtie to map reads to reference in parallel
Shuffle: aggregate alns and sort by position
Reduce: scan with SOAPsnp through the alns to find SNPs

Crossbow can run on Amazon's EMR (Elastic Map Reduce), on your own Hadoop Cluster, or on a single system.

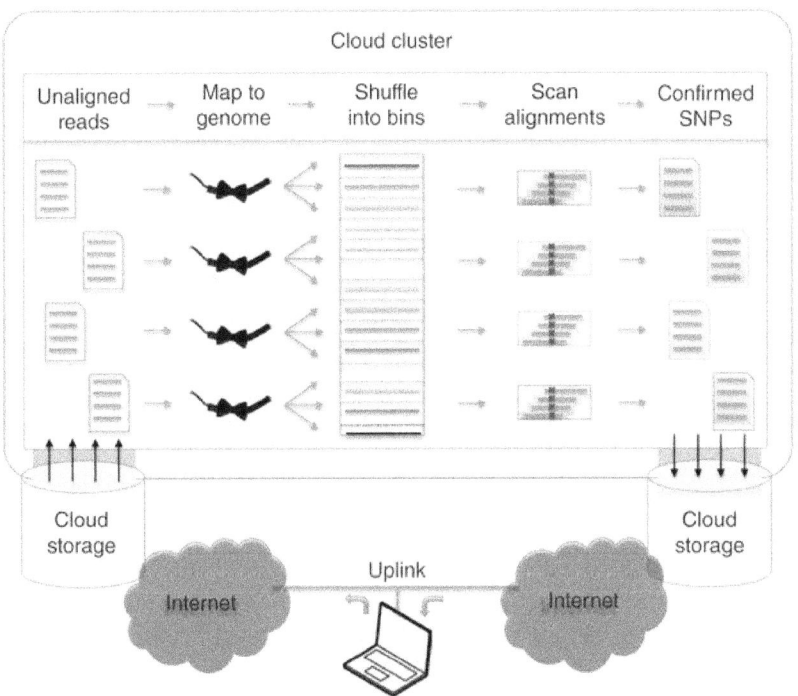

http://www.nature.com/nbt/journal/v28/n7/images_article/nbt0710-691-F1.gif

BioPig

Built on Pig
Greatly reduces parallelization time- an mpi extender took 12,000 linesof code, vs 31 in BioPig
Runs on all Hadoop
Scales well

BioPig Modules

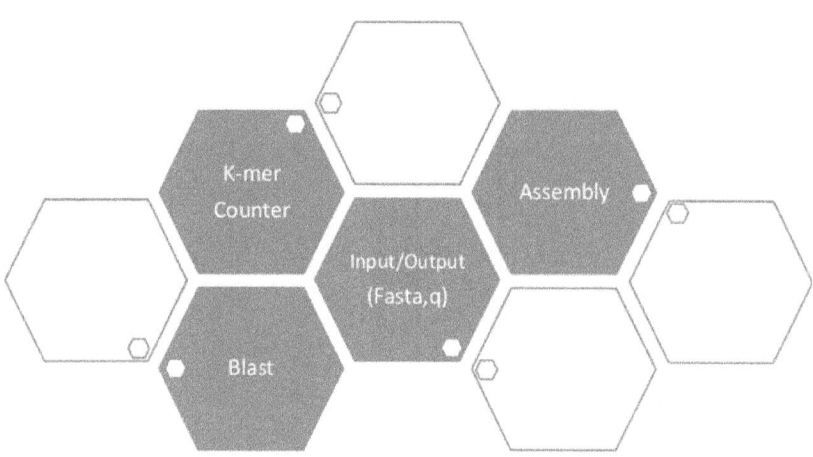

http://www.slideshare.net/ZhongWang3/biopig-for-large-sequencing-data

HBLAST

HBLAST is a Hadoop-optimized implementation of the BLAST algorithm. Both the Database and the Query data are partitioned across the cluster using Map/Reduce. The authors use a dynamic partitioning method to optimize the usage of the underlying hardware. As a result, the CPU's are efficiently utilized, even with very large clusters, making the analysis very fast.

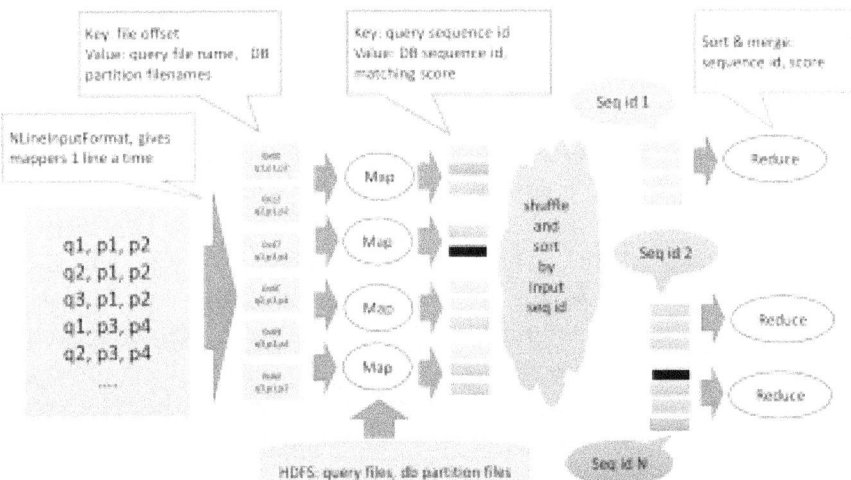

http://www.j-biomed-inform.com/article/S1532-0464(15)00010-6/abstract

ABOUT THE AUTHOR

Martin Gollery has over fifteen years of experience in bioinformatics and technology. Mr. Gollery started his career in bioinformatics computing at TimeLogic where he developed databases, technical guidance, and optimal guidelines for system performance. Mr. Gollery also served as the Director of Bioinformatics at the University of Nevada at Reno, where he coordinated over $40m in government grants and managed technicians, system administrators, and developers. While at Active Motif, a global biotechnology company dedicated to enabling epigenetics research, he led the development of the DeCypher Accelerator: algorithms that are accelerated on reconfigurable platforms such as FPGAs that can often run 100-1000 times faster than a conventional CPU. Mr. Gollery has since served as the Director of Sales and Business Development at Omnixon, which offers pre-built, scientifically validated, computationally efficient bioinformatics analysis pipelines as web-based services. Today, Mr. Gollery runs an independent bioinformatics and technology consulting practice and also serves as the Director of Sales and Marketing Support for BioDatomics, which designs intuitive software platforms for NGS data analysis.